UK Professional Pest Control Terminology

A Guide to Pest Management Reporting

UK
Professional
Pest Control
Terminology

A Guide to Pest Management Reporting

Prepared for Publication by iGuides

© 2012 iGuides

Feedback to: publications@iGuides.com.au

ISBN-13:
978-1481909099

ISBN-10:
1481909096

A

A/C: An abbreviation for air conditioner or air conditioning.

A/C Condenser- The external fan unit of the Air Conditioning system. Which removes the heat from the gas and "turns" the gas back into a liquid and pumps the liquid back to the coil in the furnace.

A/C Disconnect- The main electrical ON-OFF switch near the A/C Condenser.

Abamectins: Biologically derived organic compounds that disrupt the nervous system.

Abdomen: The last segment of an insect's body.

Abrasive: Capable of wearing away or grinding down another object.

Absolute Humidity: The amount of water vapour in the air, expressed as g/m^3.

Absorption: A process in which one substance permeates another or the process by which one substance adheres to the surface of another substance in a thin layer.

Abutment: Point of contact between two objects or parts.

Acceptable Daily Intake: ADI. Safe levels for pesticides residues.

Active Substance: The ingredient in a formulated product that makes the product work as a pesticide.

Acclimation: The process of allowing animals to become accustomed to experimental conditions.

Acclimatisation: The process by which animals become accustomed to a change in natural environmental conditions.

Acre: An area of 43,560 square feet. Originally, the area a yoke of oxen could plow in one day.

Action Threshold: The level of pest damage or activity reached that requires action to prevent damage from exceeding acceptable or tolerable levels.

Active Constituent: see active ingredient.

Active Ingredient: The part, proportion or ingredient in a pesticide which is directly responsible for the adverse effect on the target pest.

Activity: Term used to describe the presence of live termites discovered during an inspection.

Acuminate: Narrowing to a slender point.

Acute oral LD50: The quantity of an active agent or substance when taken by mouth that is required to kill 50 per cent of the animal under test, usually rodents.

Acute Toxicity: Effect of a substance such as a chemical that occurs immediately or directly upon exposure to that substance.

Adjuvant: An additive that enhances or increases the action or effectiveness of a substance. Chemical added to a pesticide formulation or tank mix to increase its effectiveness or safety.

Adventive: Not native and not fully established.

Aerosol: A fine mist of solid or liquid particles in a gas. A dispenser that holds a substance under pressure and that can be released it as a fine mist. Any tiny solid or liquid particle suspended in the air.

Aggregate: A mixture of sand and stone and a major component of concrete. Separate units gathered into a mass.

Aggregation Device: Monitors or bait stations that contain lots of termites or target organism.

Agitate: To stir, mix or move.

Air Pressure: Force that air exerts over a certain area due to it's weight or motion.

Air-dried Timber: Timber dried by exposure to air.

Alate: Having winglike extensions. The name given to a winged, reproductive in ant or termite colonies. A winged insect of a species having winged and wingless forms.

Alkaline: The opposite of acidic; having a pH greater than 7.

Allergen: Any substance that can cause an allergy. Allegens cause the body's defence systems to overreact. A general term for any substance that can cause an allergic reaction.

Allergic Effects: Harmful effects, such as skin rash or asthma.

Allergic Reaction: An over-reaction of the body's defence or immune system.

Alluvial Plain: An area of fairly flat land where a river has deposited silt.

Altruism: The quality of unselfish concern for the welfare of the colony shown by in ants and termites.

Ambient Temperature: The temperature that is characteristic of the atmosphere surrounding a small-scale feature, such as a cumulus cloud.

Ametabolous Lifecycle: The lifecycle of primitive insects such as silverfish. No changes occur to the body, the insect simply moults and grows larger. Immature and mature are similar in appearance.

Amphibious: Able to live both on land and in water.

Anafalaxis: A mild or severe reaction to an allergen.

Anchor bolt: A device for connecting timber members to concrete or masonry.

Anemometer: Instrument for measuring velocity of airflow.

Annulate: Formed in ring-like segments or with ring-like markings.

Antenna: One of a pair of mobile appendages on the head of e.g. insects and crustaceans; allowing it to sense its environment, the longer the antennae the more acute the sense of smell and movement they are aware of.

Antennal: Relating to the antenna.

Anticoagulate: The hindering of the blood clotting process especially by treatment with an anticoagulant.

Antifeedant: Any substance that prevents an insect or other pest from feeding but does not directly kill them. The target organism dies of starvation. Antifeedants are usually nontoxic.

Anti-microbial Pesticide: A pesticide used for control of microbial pests.

Appendage: Any limb or other organ which is attached to the body by a joint.

Appendages: External body parts that protrude from an organism's body.

Application Type: Pesticides applied in one of several ways: sprays, granules, mists, fogs, gels, dusts etc.

Application: The dose of pesticide which is used on a crop.

Appraisal: An expert valuation of property.

Approval: Registered or allowed, acceptance.

Approved: Judged to meet or exceed specified standards

Apron: A trim board that is installed beneath a window sill.

Aptera: Primitive wingless insect group.

Apterous: Lacking wings.

Aquaculture: The farming of fish or other aquatic organisms under artificial conditions.

Aquatic: Living in water.

Arachnid: A group of Arthropods including spiders, scorpions and ticks.

Arboreal Nest: A termite nest that is above ground level, usually in a tree.

Arboreal: Inhabiting or frequenting trees. Attached to, found in, or upon, or frequenting, trees as in some termite species.

Arch: A curved structure. Masonry construction for spanning an opening and supporting the weight above it.

Architect: Someone who creates plans to be used in making something such as homes.

Area Wells: Corrugated metal or concrete barrier walls installed around a basement window to hold back the earth

Art Deco: A decorative style popular in the 1920s and 1930s

Art Nouveau: A decorative style popular in the late 19th and early 20th centuries

Arthropod: Invertebrate having jointed limbs and a segmented body with an exoskeleton made of chitin. Spiders, insects, millipedes, centipedes and crustaceans all belong to this group.

Articulated: Consisting of segments held together by joints.

Asbestos: A non-flammable, natural mineral fibre with health-threatening properties.

Asthma: A disease in which the airways suddenly narrow making it extremely difficult to breathe. Asthma can often be triggered by an allergen.

Atrium: The central area in a building; open to the sky. A central area in a structure with a ceiling of translucent material that admits sunlight. Any chamber that is connected to other chambers or passageways.

Attic access: An opening that is placed in the ceiling of a home to provide access to the attic.

Attic Ventilators- In houses, screened openings provided to ventilate an attic space.

Attractant: Anything that attracts insects or vertebrates, sometimes used in conjunction with residual insecticides and as an essential component of insect monitoring stations.

Automatic Gates: An automatic opening gate with a motor.

Awning: A canopy made of canvas to shelter people or things from rain or sun. A roof-like shelter extending over an area.

B

Back Split: The narrowest aspect faces the road and the stairways are at the back of the house.

Backfill: The replacement of excavated earth into a trench around or against a basement or foundation wall.

Back-siphoning: The movement of liquid pesticide mixture back through the filling hose and into the water source.

Bait Shyness: The propensity for target species to avoid poisoned bait.

Bait Station: A container designed to hold bait.

Bait: A matrix containing an insecticide with a delayed toxicity and a bait or food that is attractive to the target species. The poison then kills the target species after a period of time.

Baiting: A treatment method in which a suitable bait for the pest species is strategically placed in a suitable container in order to facilitate feeding.

Balustrade: The rail, posts and vertical balusters along the edge of a stairway or elevated walkway. A railing at the side of a staircase or balcony to prevent people from falling.

Base or baseboard: A trim board placed against the wall around the room next to the floor. Skirting boards.

Batch number: A number to identify a material which is produced under uniform conditions.

Batch: A quantity of material which is produced under uniform conditions.

Batt: A section of fiber-glass or rock-wool insulation.

Batten: Narrow strips of wood used to cover joints.

Bay window: Any window space projecting outward from the walls of a building.

Beam: Structural member which supports a load.

Bearer: A sub floor timber beam placed across piers to support the floor.

Bearing Wall: A wall that supports any vertical load in addition to its own weight.

Beaufort Wind Scale: A wind scale developed by Admiral Sir Francis Beaufort in 1805 to help sailors quantify wind measurements. Values range from Force 0 (Calm) to Force 12 (Hurricane Force, >64 knots). See Section E (page 440) of the official WMO guide for more.

Bedrock: A subsurface layer of earth that is suitable to support a structure.

Beneficial insect: Any insect that is advantageous to man, by, but not limited to, preying on or competing with pests.

Berm: A narrow artificial ridge of earth along the side of a road An artificial ridge of earth.

Bevel: Any angle not at 90 degrees. Two surfaces meeting at an angle different from 90 degrees.

Bifenthrin: A pyrethroid insecticide that affects the nervous system of insects.

Bi-Level: A house built on two levels; a split-level house.

Bioaccumulation: The gradual increase in the amount of a substance in an organism.

Bioavailablity: The extent to which a pesticide residue can be taken up into an organism.

Biochemical Pesticide: A naturally-occurring substance that controls or manipulates pests by a mechanism other than toxicity.

Biochemical: A chemical that occurs naturally in an organism, involving chemical processes in living organisms.

Biocide: A chemical pesticide that is toxic to a broad array of species or can deter, render harmless, or exert a controlling effect.

Bioconcentration: The process by which the amount of pesticide or other substance is concentrated in an organism.

Biodegradation: The breakdown of the chemical structure of a pesticide by or in organisms.

Biodeterioration: The breakdown of materials by microbial action or biological agents such as fungi and insects.

Biological control: The regulation of pest populations by natural enemies.

Biological insecticides: Biological insecticides are living organisms or the toxins produced by them. A non-man made chemical derived from natural sources such as plants, fungi, bacteria. Insecticides of biological origin.

Biological pesticide: Biological pesticides are living organisms or the toxins produced by them. A non-man made chemical derived from natural sources such as plants, fungi, bacteria.

Biomass: The weight of living material in a particular section or segment of habitat or area.

Biopesticides: Biological pesticides are living organisms or the toxins produced by them. A non-man made chemical derived from natural sources such as plants, fungi, bacteria. Pesticides of biological origin such as pyrethrum.

Biota: All living organisms. All the plant and animal life of a particular region.

Biotic: A descriptor of the living components of ecosystems.

Bird Repellent: Any chemical, substance or method that deters birds from approaching or feeding in a certain area.

Bivouac*:* A termite outpost or re-hydration structure where termite workers and soldiers re-hydrate before heading back to the nest.

Blow insulation: Fibre insulation in loose form and used to insulate attics and existing walls where framing members are not exposed.

Blue print: Photographic print of plans or technical drawings etc.

Board Foot: (fbm) 144 cubic inches (12 in. × 12 in. × 1 in.). Used for lumber.

Bolt: 40 yards. Used for measuring cloth.

Boric Acid: Boric acid and its sodium salts are used as insecticides, some act as stomach poisons in insects such as ants, cockroaches and termites, while others are used because they scour the body covering exoskeletons of insects.

Botanical Pesticides: Insect toxins that are derived or extracted from plants or plant parts.

Bottom plate: 2 x 4's that lay on the subfloor upon which the vertical studs are installed.
Boundary Fence: A perimeter fence along the boundary of a property.

Brace: A structural member used to stiffen a framework.

Brachypterous: Having very short or rudimentary wings or not fully developed.

Bracing: A structural member used to stiffen a framework.

Breakdown Product: When some chemicals enter the environment they break down into other chemicals. These new chemicals are called breakdown products.

Brick veneer: A vertical facing of brick.

Broadcast Spraying: An imprecise, traditional method of dispensing chemical pesticides over a large area, broadcast spraying is prone to pesticide drift.

Brood: A group of offspring.

Bubonic plague: A highly contagious, deadly disease caused by the bacteria 'yersinia pestis' (formerly Pasteurella pestis) and transmitted by the fleas of rodents.

Budding: Formation of a new colony by the splitting of an existing colony.

Building Code: An ordinance or regulation designed to control the design, construction or alteration and repair of buildings and structures which can be enforced by local councils.

Building codes: Community ordinances governing the manner in which a home may be constructed or modified.

Built-Ins: Items meant to be permanent fixtures in a building which, if taken away, would leave evidence of there removal such as cabinets, benches and shelving.

Bull nose: Rounded corners.

Bungalow: A one floor building with a basement.

By fold door: Doors that are hinged in the middle for opening in a smaller area than standard swing doors. Often used for closet doors.

C

Cache: A protected source of bait. As in above ground stations.

Calibration: The act of checking or adjusting (by comparison with a standard) the accuracy of a measuring instrument.

Calm: A state of no wind. Force 0 on the Beaufort Scale.

Cannibalism: The eating individuals of the same species.

Cantilever: An overhang. Projecting horizontal beam fixed at one end only.

Cap flashing: The portion of the flashing attached to a vertical surface to prevent water from migrating behind the base flashing.

Cap: The top part of a structural member or the upper member of a column, door cornice, molding, or fireplace.

Capacity: The water handling capability of a pump expressed as gallons per minute (GPM).

Centrifugal force: The action that causes something to move away from its center of rotation.

Carbamates: A salt, or ester, of carbamic acid. Carbamate as insecticides have toxic mechanisms similar to that of organophosphates, but have a much shorter duration of action and are thus considered less toxic.

Cape Cod: Storey and a half home with two or more dormer windows in the roof.

Carbaryl: A white crystalline, solid, chemical in the carbamate family used as an insecticide.

Carcinogenic: The ability of a substance to cause or tend to cause cancers.

Carnivore*:* Any animal that feeds on flesh. Animals that kills and consume other animals.

Carport: A roofed auto shelter without walls.

Carrier: The primary material used to allow a pesticide to be dispersed effectively. A substance which can be mixed with an active ingredient to assist application. Can also be used to reduce the strength of a product to a desired level.

Casement Window: A window with hinges on one of the vertical sides so it can swing open.

Casement: Frames of wood or metal enclosing a window sash.

Casing: Wood trim border installed around a door or window.

Caste*: In some insects a physically distinct individual or group of individuals specialized to perform certain functions in the colony.

Cathedral ceiling: This is a high sloping ceiling that is attached directly to the roof trusses and rises all the way to the top of the house.

Caudal: Relating to the tail, or tail end, of an insect.

Caulking: A waterproof filler and sealant.

Cca: Copper chrome arsenate, a wood preservative.

Ceiling joist: One of a series of parallel framing members used to support floors or roofs.

Cellar: A storage space below ground level in a structure.

Cellulose: A polysaccharide that is the chief constituent of all plant tissues and fibres. About 33% of all plant matter is cellulose. Humans can digest cellulose to some extent, however it is often referred to as 'dietary fiber' or 'roughage' and acts as a hydrophilic bulking agent for faeces.

Celsius (C): A temperature scale where water freezes at 0º C and boils at 100º C.

Cement: A building material that is a powder made of a mixture of calcined limestone and clay; used with water and sand or gravel to make concrete and mortar.

Central nest: The main colony where reproduction occurs.

Century Home: Home erected over one hundred years ago.

Cephalic: Relating to the head.

Cerci: Paired appendages on the rear-most segments, tip of the abdomen, of many arthropods, including insects and arachnids. Cerci are sensory organs.

Cesspool: An underground catch basin for household sewage or liquid waste.

Chainmesh Fence: A specific type of perimeter barrier, often constructed of steel with see through rings.

Chelicerae: The jaws or fangs of an arachnid. Often modified for grasping and piercing.

Chemical name: Scientific name for the active ingredient.

Chemical: Material produced by or used in a reaction involving changes in atoms or molecules.

Chemosterilant: A chemical that renders an insect infertile thus preventing it from reproducing.

Chip Board: An inexpensive hard material made from wood chips that have been pressed together and bound with synthetic resin.

Chironomids: Midges found in aquatic environments. This is a large group of insects with over 5000 described species.

Chitin: A tough semitransparent substance, the hard outer shell of an insect.; the principal component of the exoskeletons of arthropods. Also called the cuticle.

Chlorfluazuron: An insect growth regulator (IGR) also known as a chitin inhibitor. Affecting the moulting process of termites, chlorfluazuron in the cellulose bait ensures the nymph cannot produce enough chitin to form a new cuticle and so dies. As the nest is humid and warm the decomposition of dead termites combined with fungal growth, that the remaining workers cannot clear, causes the whole nest to collapse and die including the queen. Used in subterranean termite baiting stations

Chronic Effect: The effects of exposure to a pesticide over a long period of time.

Chronic Exposure: Continued exposure over a lengthy period of time.

Chronic toxicity: Effects of a substance that occur as a result of numerous exposures to sub-acute toxicity doses. The capability of a pesticide to produce injury following chronic exposure, immediately upon exposure, or delayed

Cimex Lectularius: Parasitic insects. The most common type is *Cimex lectularius*. All insects in this family live by feeding exclusively on the blood of warm-blooded animals.

Cladding: The external protective covering on the outside of a building. Can be natural, treated timber or plywood.

Class: A division of the animal kingdom lower than a phylum and higher than an order. A taxonomic group containing one or more orders.

Climate: The average state of the various atmospheric variables of temperature, precipitation, wind, etc over a period of time.

Cloud: A visible collection of very fine water droplets or ice crystals suspended in the atmosphere.

Coat: A single layer of paint, plaster or other material.

Cocoon: A protective bag of silk like substance spun by larvae to protect pupas or by spiders to protect their eggs.

Cold Front: A frontal boundary where a cold airmass is replacing warmer airmass.

Coleopterans: Species commonly called Beetles. Approximately 40% of all described insect species are beetles, about 400,000 species.

Colonial: Large home with pillared entrances.

Colony: A group of social organisms of the same type living or growing together.

Colorbond Steel Fencing: A particular brand of steel fences, which are made from Colorbond steel or similar.

Column: A tall vertical cylindrical structure standing upright and used to support a structure.

Commercial Buildings: Structures designed and used for business purposes.

Common Area: An area that is owned, used or

Communicate: The provision of information by chemical, touch, or noise.

Community: All organisms inhabiting a common environment and interacting with one another.

Complete Lifecycle: Organisms that pass through four distinct stages – the egg, larva, pupa and adult. Newly hatched young look like grubs or caterpillars. See complete lifecycle. Also known as a Complete Metamorphosis.

Complete Metamorphosis: Development where there are four life stages of egg, larva, pupa and adult. Also known as a Complete Lifecycle.

Compliance: Conforming to official requirements.

Compound eye: Eye made up of separate, light-sensitive visual elements, lenses, known as ommatidia, each forming a portion of an image. Compound eyes are extremely sensitive to motion. Some arthropods have compound eyes of only a few ommatidia, while some have many, each being able to create a separate image. With each eye viewing a different area or object, a coalesced image from all eyes is created in the brain, often providing high-resolution images.

Compound: Put or add together. A substance composed of two or more chemicals. Make stronger.

Concentrated: Relating to a solution whose dilution has been reduced. Having a high percentage of active ingredient.

Concentrates: Having a high percentage of active ingredients.

Concentration: The strength of a solution. The amount of active ingredients in a given formulation.

Condensation: The process by which vapor changes to a liquid. Increasing the density of something

Condominium: A structure where the interior spaces are individually owned but the balance of the property, both land and building, is owned in common by the owners of the individual units. The balance of the property is called the common area.

Conduction: The direct transfer of heat energy through a material.

Conductivity: The rate at which heat is transmitted through a material.

Conduit, electrical: A passage through which electric wires can pass.

Conservation: The preservation, management, protection and sustainable use of natural resources over the long term.

Construction, frame: Construction type in which the structural portions depend upon a frame for support.

Contact pesticide: Pesticides that penetrate the target pest's body wall. Contact pesticide is usually absorbed through the skin or outer membrane. The pest is affected when directly exposed to the product.

Contagious: A disease that is capable of being transmitted to another person.

Contaminate: Make impure in a bad way; make something harmful, especially by the addition of some unwanted substance.

Convection: Currents created by heating air, which then rises and pulls cooler air behind it.

Conventional pesticide: Man made chemicals that can be used to kill pests

Conveyancer: A professional who assists in the legal transfer of property.

Conveyancing: Act of transferring property title from one person to another.

Cooling-off period: The period when a person may legally withdraw from a contract without incurring a penalty.

Coprophagy: Eating of faeces.

Copularium: Another name for the nuptial chamber.

Corbel: Timber laid horizontally on the top of a column to transfer loads.

Coriaceous: Leather-like in texture, tough but pliable.

Cornice: A moulding used at the join of a ceiling and a wall.

Cosmopolitan: Occurring throughout most of the world.

Country Property: A rural property surrounded by fields.

Cove ceilings: Rounded concave surface joining the walls and the flat ceiling.

Coxa: The basal segment of the insect leg.

Crack and Crevice: Application of insecticides or the inspection in structures to cracks and crevices where pests may live and breed.

Crawl space: A shallow space below a house.

Crepuscular: Active in the twilight.

Cruciform: In the shape of a cross.

Cryptic: Favouring concealment.

Cubit: 18 inches or 45.72 cm. Average distance between elbow and tip of middle finger.

Cumulative Effect: Increasing by successive addition.

Cuticle: Hard outer covering of organisms such as arthropods secreted by the epidermis. The principal component of the exoskeletons of arthropods. Also called chitin.

Cuticular: Pertaining to the cuticle.

Cutworms: see Bogong Moths. See Common Pest Section.

Cyclone Wire Fence: A specific type of perimeter barrier, often constructed of steel with see through rings.

Cyfluthrin: Cyfluthrin belongs to a class of insecticides known as synthetic pyrethroids. Synthetic pyrethroids are man-made and copy the properties of the naturally-occurring insecticide pyrethrum, which comes from the Chrysanthemum flower. Synthetic pyrethroids are believed to have longer residual action than pyrethrums.

D

Damaged timber: Wood that has been spoiled by termite activity, wood decay etc.

Damp Proofing: Any treatment that prevents the passage of moisture.

Dealate: An insect that has naturally lost its wings. Ants and termites shed their wings after a mating.

Decay: The organic process of rotting. Bacterial decay, decomposition of organic matter.

Decibel: A logarithmic unit of relative loudness. One decibel is the smallest amount of change detectable by the human ear. A logarithmic unit of sound intensity.

Deck: Raised, uncovered outdoor area normally constructed with treated timber with a gap between the flooring

Decontaminate: To remove a contaminant. Remove pesticide so as to reduce risk or damage.

Defoliation: The loss of foliage. The process when a plant looses all its leaves.

Degradate: The chemical product resulting from the breakdown of a pesticide.

Degradation: Changing to a lower state. A pesticide which is broken down to simpler substances. Also known as decomposition.

Degrade: Lower the grade of something. Reduce in worth. The result of any process that lowers its value as in wood.

Delayed effects: Effects, illnesses or injuries that do not appear immediately.

Dengue fever: An infectious disease transmitted by mosquitoes and characterized by a fever, rash, and severe pain in the joints.

Density: The mass per unit volume of an object.

Dermal toxicity: The toxicity of a chemical associated with absorption through the skin or outer membrane.

Desiccant: A substance that promotes drying through the loss of water. Desiccants kill by removing or disrupting the outer coating on the insect's cuticle causing loss of body fluids resulting in the insect dying of desiccation or dehydration.

Desorption: Changing from an adsorbed state on a surface to a gaseous or liquid state. The release of a substance through the surface of an organism.

Deterioration: Reduced quality or strength.

Detoxification: Neutralizing toxic properties.

Detritus: Organic debris from decomposing material.

Dew: Water that has condensed on a cool surface overnight from water vapour in the air. Frost is a frozen form of dew.

Dewpoint: The temperature to which air must cool for condensation to occur. The air must be saturated before condensation begins

Diapause: An insect resting stage, a temporary pause in the growth and development, usually induced by environmental signals or extreme conditions like winter or summer.

Diatomaceous earth:. A light, naturally occurring soil consisting of siliceous diatom remains. It is soft and easily crumbled into a fine powder. In some insects the coarse texture abrades the outer waxy coating that keeps water inside and desiccates some insects.

Diluent: A diluting agent which reduces the concentration of a formulation to the desired strength.

Dilute: To make less concentrated.

Dimorphic: Occurring in two distinct forms.

Dimorphism: The existence of two forms of individual within the same animal species but not related to differences in sex.

Diptera: Derived from the Greek words "di" meaning two and "ptera" meaning wings. Refers to a large order of insects having a single pair of wings and sucking or piercing mouths such as mosquitoes, gnats and and true flies.

Dissipation: Breaking up and scattering by dispersion.

Diurnal: Active in daytime.

Domestic Buildings: Structures designed for families.

Doorjamb: The surrounding case into which door opens of closes.

Dorsal: Belonging to the upper surface of an organism or part.

Dose: The quantity of an substance required to achieve its aim.

Double-Pitch Roof: A roof that slopes in two directions.

Downspout: A pipe for draining water from roof gutters.

Drainage: The rate that water will pass through soil or a system of drains

Dressed timber: Dressed timber has been planed to a smooth finish on one or more surfaces.

Drift: Movement of pesticide in the air and away from the recognized treatment area.

Drone: Stingless male bee whose exclusive purpose is to mate with the queen.

Drosophila melanogaster: see Common Vinegar Fly. See Common Pest Section.

Dry rot: The decay of timber in buildings and other wooden structures caused by a certain fungi.

Ducts: Individual or multiple passages or tubes used for conveying air as in a ducted air conditioning system.

Duplex: A house containing two separate dwelling units, side by side or above the other.

Dust: A dry formulation of an insecticide that attaches to an insect when the insect walks over it.

Dusting: A treatment technique for applying a thin coat of dust to an area.

Dwarf Wall: A wall of less height than a full story of a building also a wall which supports the sleeper

joists under the lowest floor of a building. i.e.
Dwarf walls in subfloor areas or a structure.

E

Eaves: The horizontal overhang at the lower edge of a roof.

EC: Emulsifiable concentrates.

Ecdysis: See Moult.

Ecology: The environment as it relates to living organisms.

Ecosystem: A system formed by the interaction of a community of organisms with their physical environment.

Ectoparasite: A parasite that lives on the outside of its host. Lice and fleas are ectoparasites.

Egress: A means of exiting a structure.

Elevated Ranch: Similar to ranch home, but basement at ground level.

Elytra: Hard forewings of some insects such as beetles which cover and protect the functional hind flight wings.

Embedment; Embedded: Set firmly into a surround mass.

Emulsifier: substance that allows two liquids to mix which would not normally mix. Chemical that allows petroleum-based pesticides like emulsifiable concentrates (EC's) to mix with water.

Emulsion: A mixture of two or more liquids that are not soluble in one another.

Encapsulation: The condition of being enclosed. The surrounding of an invading body by insect blood cells and the formation of a protective outer casing.

Endangered species: Flora or fauna in imminent danger of extinction.

Endemic: Restricted to a well defined geographical region.

Endoskeleton: The internal skeleton; bony and cartilaginous structure.

Entomologist: A person who studies insects.

Entomology: The study of insects.

Entomopathogenic: Insect-attacking organism.

Environment: The totality of surrounding conditions.

EPA: Environmental Protection Agency.

Ephemeral: Organisms with short life cycles, usually adapted to making rapid use of favourable environmental conditions.

Epidemic: Occurs when a certain disease exceeds what is expected of it based on recent experience.

Eradicate: Destroy completely.

Ester: Formed by reaction between an acid and an alcohol with elimination of water.

Estimate: The cash amount that a contractor anticipates spending on materials and labour during the course of a project.

Evaporation: The process by which a liquid changes to a gas or vapor.

Exclusion Treatment Method: Protecting a given area from a target species by physical exclusion methods such as barriers to access points and by the use of pesticide repellence, internal and external barrier treatments, pesticide exclusion. Exclusion involves altering the environment so pests can not get through, or into, the exclusion zone

Excrement: Waste matter such as urine or sweat but especially faeces discharged from the body.

Exoskeleton: Collectively the external sclerites of the integument. The hard outer covering of all Arthropods.

Expansive soils: Earth that swells and contracts depending on the amount of water that is present.

Explanate: Spread out and flattened.

Exploitation competition: Competition between species where one suppresses another's rate of increase by use of a limiting resource.

Exposure: The state of being vulnerable or exposed to a pesticide.

Exterior Finish: The outside protective layer.

Exterior Fixture: An outside item, e.g., areaway, canopy, marquee, platform, loading dock, that is permanently attached to the building structure.

Exterior Wall: An outer wall.

Exterminator: Old term for a pest control technician.

Exuvia: The cast-off outer skin of an insect or other arthropod.

Eyes: The Organs of sight.

F

Façade: The principal, exterior face of a structure.

Face: The most important side of a structure.

Faeces: Solid excretory product evacuated from the bowels through the anus.

Family Room: An informal living room.

Fascia: Horizontal boards attached to truss ends at the eaves and along gables.

Fauna: All the animal life in a particular region or period.

Feral animals: Introduced animals that have reverted from domestication to a wild state (e.g. horses and goats)

Feral: Wild and menacing. Any domesticated species that has reverted to a wild state.

Filament: A thread-like structure.

Fittings: Items that can be removed from a property without damage.

Fixtures: Anything permanently attached to a house and regarded as part of the real property, such as cabinets and cupboards.

Flagstone: A flat, slab of stone used for paving walks, patios and terraces.

Flange: Projecting rim or collar that increases stiffness.

Flashing: A strip of impervious metal shaped and attached to a roof for strength and weatherproofing.

Flashing: Waterproof material that prevents moisture from penetrating a house through the walls or roof.

Flat: A self-contained dwelling unit in a multi-unit building.

Flooring: The covering of internal floors in a building.

Flora: All the plant life in a particular region or period

Flow rate: The mass of substance which passes through a given surface or piece of equipment. Also known as volume flow rate or rate of fluid flow.

Flushing Agent: Any substance that forces insects to leave an environment they consider safe.

Fog: condensed water vapor or a cloud floating close to the ground.

Foliaceous: Resembling a leaf.

Fontanelle: Any membranous gap between the bones of the cranium, as in some species of termites such as Coptotermes frenchi.

Food chain: Predator-prey relationships between species within an ecosystem or habitat.

Foraging: To hunt or search for food. The act of searching for suitable food for an individual or a colony.

Forceps: Pinchers at end of body.

Fore Wings: The first pair of wings on an insect's body.

Forecast: A prediction of the weather into the future.

Foreleg: A front leg.

Forewing: Either of the two front wings of a four-winged insect.

Form Work: Temporary structure, of wooden frame, erected to contain concrete.

Formulation: A mixture of active and inert ingredients.

Fossorial: Adapted for digging.

Foundation: The base on which something is supported where all loads are transferred to the ground.

Frame: The main internal timbers of a structure that are fitted and joined together and give a structure its shape.

Framing timbers: Timbers that are used to form the basic structure of a building.

Frass: The material, refuse or excrement left by certain pests and usually a mixture of faecal and plant matter.

Front Fence: A fence along the front of a property.

Full brick: A building where both the inside and outside walls are brick.

Fumigant: Pesticide that is applied as a vapour or gas that enters the pest's body via inhalation or through the spiracles.

Fumigants: Pesticides in the form of a gas

Fumigation: The application of a gas or smoke to eradicate pests.

Fungus: An organism of the kingdom Fungi lacking chlorophyll and feeding on organic matter; ranging from unicellular or multicellular organisms to spore-bearing syncytia.

G

Gable roof: Has a triangle on the side or front of the façade.

Gable Vents: A louver vent mounted in the top of the gable to allow air to pass through a roof void.

Gable: The vertical triangular wall between the sloping ends of gable.

Gallery: A tunnel or void created by termites in soil or timber.

Galley kitchen: A smaller than usual kitchen with cupboards usually arranged along a single wall.

Gastropods: Molluscs including snails and limpets.

Gel Bait: A formulation of a bait contained within a gel like substance that acts as a food source.

Generalist: A pest that can use a wide range of species as host or prey.

Genitalia: External sex organ.

Genotype: Genetic makeup of an individual.

Genus: A general kind of something eg Coptotermes, Nasutitermes and Schedorhinotermes.

Geodesic: The shortest line between two points.

Girder: A large beam of wood or steel used to support loads.

Glass Pool Fencing: A pool fence or perimeter barrier made from safety glass.

Grading: Sloping of the grounds immediately adjacent to a building helping water flow away from a building or structure.

Gradual Metamorphosis: Development in which there are three life stages of egg, nymph and adult where the nymphs resemble the adults.

Grain: The direction of fibres in wood.

Gram: One thousandth of a kilogram

Granular Bait: A granule formulation of a bait contained within a substance that acts as a food source.

Gray Water: Waste water not containing sewage or fecal matter or food wastes.

Great Gross: 12 gross or 1728.

Green Lumber: Unseasoned timber. Freshly cut lumber that has not had time to dry. **Green**

Products: Those products that have minimal or no impact on the environment or to human health.

Green Timber: See Green Lumber.

Grooming: Social behaviour where termites clean each other.

Gross: 12 dozen or 144.

Ground Cover: Low growing plants which create a foliage blanket over an area.

Ground water: Water beneath the earth's surface in soil or rock.

Gut symbionts: Organisms that live in the gut aiding in the digestion of food. In termites, these are commonly protozoa, fungi and bacteria. See protozoa.

Gutter: A shallow channel set below the lower edge of a roof and along the fascia eaves to catch rainwater from the roof.

H

Habitat Manipulation: Manipulation of the environment to minimise the effects of natural enemies. Changing an environment so it is less suitable to target species making sure food, water and harborage are not made available.

Habitat Modification: See Habitat Manipulation.

Habitat: The type of typical environment in which an organism or group naturally lives or occurs.

Haemolymph: The fluid that circulates within a termite.

Half-Life: Time taken for a pesticides strength to decline by half.

Hall: A room at the entrance of a building that provides access to other areas of a structure.

Hantavirus Pulmonary Syndrome: A usually fatal disease transmitted by rodents. A third of people contracting Hantavirus will die. Inhalation of urine or faecal contaminated dust is thought to be the main form of contamination of humans.

Harbourage: Safe refuge living areas for pests. Any shelter or place of refuge that protects pests from outside influences.

Hardwood: Wood harvested from broadleaf trees such oaks, maples, ashes and elms.

Hazard Assessment: Factors controlling the possible effects of a hazard.

Hazard: A source of danger.

Heartwood: More durable wood from the centre of a tree.

Heat stress: Heat stroke occurs when a person becomes dehydrated and their body temperature rises above 40.5C. This is a medical emergency and can lead to death.

Hemimetabolous: Changing gradually from larva to adult with no pupal stage. An incomplete metamorphosis.

Herbacous: Dying down at the end of the growing season.

Herbicide: A pesticide used to kill or control undesirable plants and weeds.

Herbivore: Any animal or organism that feeds on grass and other plants.

Herbivorous: Feeding on plants.

Hexapod: Animal with six legs.
High Toxicity: A characteristic of a chemical or substance used in pest control that that make them a high risk of poisoning to non target animals.
Hind Wings: The second pair of wings on an insect's body.

Hindleg: A back leg.

Hindwing: Either of the posterior wings, borne on the metathorax.

Hindwings: The second pair of wings of an insect.

Hip and valley roof: Used for L or U shaped buildings. The valley is formed at the internal junction of the two roofs.

Hipped Roof: An even roof to wall junction for the whole house and eaves on all sides.

Histoplasmosis: A disease transmitted by birds. This disease affects the lungs and can be fatal if untreated.

Holometabolous: Undergoing complete metamorphosis with larval and pupal stages.

Honeydew: A sweet liquid excreted by plant-sucking insects such as aphids.

Horsepower: The power needed to lift 33,000 pounds a distance of one foot in one minute.

Host: An animal or plant that nourishes and supports a parasite from which it does not benefit. Sometimes the host is actually harmed by this association. For example the host of a flea is usually a dog.

Humidity: A measure of the amount of water vapour in the air.

Hybrid: Individual resulting from interbreeding of two species.

Hydrophilic: Having a strong affinity for water. Tending to dissolve in or mix with water.

Hymenoptera: An order of insects including bees, wasps and ants. Hymenoptera is one of the largest insect orders and there are over 120,000 recognized species.

Hypostome: The harpoon-like structure on the head of a tick that allows them to anchor themselves to their host during feeding.

I

IGR: Insect Growth Regulators / Inhibitors. Insect growth regulators interrupt or inhibit the life cycle of a pest. A substance that controls or modifies insect growth.

Imago: The adult insect.

Immature: Not yet fully developed.

Incomplete Lifecycle: Lifecycle of an organism where the immature organism closely resembles the adult and simply increases in size with each moult.

Incrassate: Thickened.

Index of abundance: An indicator of relative number or density of a species.

Indicator organisms: Used to measure potential faecal contamination of environmental samples.

Indigenous: Native to an area.

Inert Ingredient: Having only a limited ability to react chemically or chemically inactive. The part of a pesticide that has, or produces, no pesticidal activity.

Inert ingredients: Inactive components of a pesticide formulation. Used to dilute a pesticide.

Infest: Invade in great numbers.

Infestation: The establishment of insects in an area usually free off insects. The state of being invaded or overrun by insects, rodents or other pests within a building or area.

Inoculative release: The release of small numbers of natural enemies so they can reproduce and spread throughout an area.

Inorganic: Not forming part of the substance of living bodies. Lacking the properties and characteristics of living organisms.

Inquiline: An organism that lives within the nest of a social insect.

Insect Growth Regulator: See IGR.

Insect: A Small air-breathing arthropod in the biological family of insecta; insects have six legs.

Insecta: The Class of animals to which termites belong. Approxamately five-sixths of all known animal species.

Insectavorous: Depending on insects as food.

Insecticide: Any chemical specifically used to kill insects.

Insoluble: Does not dissolve in liquid.

Installed: Attached or connected.

Instar: An insect or other arthropod between moults.

Integrated Pest Management: A co-ordinated approach to pest management that relies on a combination of practices. It is not a single pest control method but, rather, a series of pest management evaluations, decisions and controls.

Integument: The outer layer of the insect.

Interference competition: Competition between species where one suppresses another's rate of increase by interfering with its ability to procure or by its use of a limiting resource.

Introduced animals: Animals that are not indigenous to Australia

Introduced species: Animals that are not indigenous to Australia

Intrusive Inspection: Going beyond the obvious visual indicators to also inspect areas that are not usually visible ie. behind walls, under insulation batts and other concealed voids throughout a structure.

Invasive Inspection: Going beyond the obvious visual indicators to also inspect areas that are not usually visible ie. behind walls, under insulation batts and other concealed voids throughout a structure.

Invasive animals: Wild animals that have a negative impact on the environment, agriculture or human activity

Invertebrate: Lacking a backbone or spinal column.

J

Jamb: The side of a window or door opening.
Joist: Timber beams or beams used to support floors or roofs.

Juvenile Hormone: Prevents larvae from developing into adults. A biochemical that occurs in insects and regulates their development.

K

Kelvin (K) scale: A temperature scale where 0º K represents absolute zero, the freezing point of water is 273º K, and the boiling point of water is 373º K.

Kelvin: The unit of temperature on the Kelvin Temperature Scale. 1K = 1°C.

Key Pest: An insect that frequently results in unacceptable damage or physical discomfort and so requires a control action.

Kick Plate: A metal strip placed at the lower edge of a door to protect the finish.

Kinetic Energy: The energy of motion.

King: A functional male in a termite colony.

Knockdown: A products ability to negatively affect a pest.

Knot: A unit of speed equal to 1 nautical mile per hour, 1.15 statute miles per hour, or 1.84 kilometers per hour.

L

Label: A brief description given for purposes of identification includes information relating to the active ingredient, offers instructions for product use and lists additional information, users are legally required to follow, as required by the relevant registration authority.

Labial: Concerning the labium.

Labium: A liplike structure that bounds a bodily orifice.

Laminating: Bonding together two or more layers of materials.

Landing: A platform between flights of stairs.

Landscaping Timbers: Timbers used as borders around a garden.

Larva: A young insect which is different from the adult

Larvae: The immature free-living form of most invertebrates. The immature stage of an insect between egg and pupal stages.

Larvicide: An insecticide that kills insect larvae.

Lateral Force: A force applied horizontally to a structure such as wind.

Lateral: Concerning the sides.

Lattice: A screen of crossed strips, of wood or metal.

Leaching: The movement, or gradual penetration, of a pesticide through soil or other medium.

Lead: A tunnel created by termites to protect them from the elements and to control the internal atmosphere.

Lepidoptera: Order including moths and butterflies and commonly characterized as being covered in scales, having two large compound eyes and an elongated mouthpart called a proboscis.

Liability: Legal responsibility.

Lignin: A substance produced by woody plants that binds cellulose in cell walls. The chief constituent of wood other than carbohydrates.

Limestone: Calcium carbonate, a sedimentary rock that when ground up can be used as a insecticidal desiccant.

Lineal metre: A unit of measurement in a straight unbroken line.

Link Homes: Single family dwellings linked underground at the foundations.

Lintel: A horizontal structural member that supports the load over an opening such as a door or window.

Liquid Bait: An insecticide held within an attractive bait matrix that is unknowingly ingested by pests as a food source.

Litre: Basic unit of volume in the metric system. 1000 millilitres equals 1 litre.

Litter: A group of young mammals.

Load-bearing Wall: A wall that supports weight from a floor or ceiling above it.

Louvre: A slat or fin over an opening that is pitched to keep out sunlight.

Lyme Disease: Caused by bacteria spread by the bite of some infected ticks.

M

Macroinvertebrate: An animal without a backbone and visible to the naked eye for example, worms, insects.

Macropterous: Having large wings.

Maggot: The immature form of a fly or wasp.

Malaria: Translation 'bad air', is a tropical disease spread by mosquitoes and kills over one million people a year.

Mammal Repellent: A chemical that deters mammals from an area. Usually used to deter mammals from destroying stored goods or crops.

Mandible: The laterally articulated jaws of an insect.

Mandibulate: Having a mouth with chewing mouthparts.

Mansard Home: A one storey or split level home with a low slung roof.

Marsupial: Mammals of which the females have a pouch (the marsupium) containing the teats where the young are fed and carried.

Masonry: Stone, brick or concrete.

Mass: The amount of matter in an object.

Mate: To reproduce.

Mating Disrupter: A chemical that interferes with the way that male and female insects locate each other in order to mate using pheromones.

Mechanical Pest Control: Control of pests by physical means such as the use of screens or covers.

Metallic Pony Ant: see Green Head Ant. See Common Pest Section.

Metamorphosis: The transformation of a larva into an adult that occurs in some animals.

Metathorax: The 3rd and last segment of the thorax.

Microbial: A microscopic organism.

Microceroterme turneri: Found between Port Macquarie and Townsville with a mature length of approximately 5.3 mm. This speice of Microcerotermes is an arboreal nest builder.

Microencapsulation: Micro-encapsulation is a process in which tiny particles or droplets are surrounded by a coating, protecting the efficacy of the core substance for controlled release as each capsule is burst. A microencapsulated (micro encapsulated) product can give you a longer residual than a conventional liquid concentrate

(EC) without leaving an unsightly residue found when using wettable powders.

Minimum Risk Pesticides: Products that do not pose any risk to humans or the environment.

Miscible: Capable of being mixed.

Mite: Any of several minute invertebrates belonging to the phylum Arthropoda, class Arachnida, often infesting animals, plants or stored foods.

Miticide: An acaricide that is used to kill mites.

Molar: Grinding surface of mandible.

Molluscicide: A pesticide that is used to kill slugs and snails.

Moniliform: Antennae composed of bead-like segments.

Monitoring: Regular, ongoing inspections to a site to determine effectiveness of treatment or levels of infestation or activity.

Monoculture: Domain of a single culture. An environment that has a single dominant species of animal or plant.

Monoecious: Having both male and female sex organs.

Monogamous: An animal that has only one mate.

Monolithic Slab: A continuous concrete pour is used to create the floor surface and foundation walls. This makes the foundation very strong and also protects against termite attack.

Mortar: A mixture of cement with sand and water used in masonry work.

Moult: To shed the outer covering of the body in order to grow larger. The process of shedding the exoskeleton. This is also known as ecdysis.

Mound: A typically constructed termite nest that rises above the soil surface.

Mouthparts: Structures or appendages near the mouth, adapted for use in gathering or eating food.

MSDS: Material Safety Data Sheets. MSDS contain details of the hazards associated with a given chemicals and there safe use.

Mudding: A protective shield to protect termites from the elements, predatory organisms and to control the internal atmosphere.

Mulch: A protective covering of rotting vegetable matter used to cover soil for moisture conservation and weed suppression.

Murine typhus: A disease transmitted from rats to humans by fleas, the symptoms being fever, headache, and muscular pain.

Mutagenic: Capable of inducing mutation. The ability of a substance to produce genetic changes in living cells and/or organisms.

Mycoplasmas: Any of a group of small parasitic bacteria that lack cell walls and can survive without oxygen. The smallest known living organisms.

N

Native animals: Animals that are indigenous to Australia

Native species: Animals that are indigenous to Australia

Natural Pest Management: Pest Management treatments that emply low-impact natural, organic, botanical and biological material to control pests.

Natural Pesticides: Products derived from naturally occurring substances such as plants.

Nematodes: Small unsegmented worms with elongated rounded body pointed at both ends. Mostly free-living but some are parasitic.

Neonicotinoid Insecticides: A class of insecticides which act on the central nervous system of insects with lower toxicity to mammals. Neonicotinoids are synthetic analogues of the natural insecticide nicotine.

Nocturnal: Active at night.

Nogging: A short piece of lumber set between two studs or joists to keep them rigid.

Non-bearing Wall: A wall supporting no load other than its own weight.

Non-Expansive Soil: A soil which does not experience considerable volume changes due to changes in moisture content.

Nonporous surfaces: Surfaces that have no openings or pores that would allow a liquid to be absorbed.

Non-target animals: Animals co-existing in the same environment but are not subject to management.

Non-target Organism: Any living organism that a pesticide is not intended to control.

Non-Target Species: Species that are not the primary target of a pest control treatment or plan but run the risk of being killed or harmed.

Nymph: An insect with incomplete metamorphosis. An immature stage in an insects life cycle.

O

Obligatory: Not optional.

Obtuse: Not pointed or acute.

Ocellus: One of the simple eyes of insects.

OCP: Organochlorine pesticide. Pesticides containing chlorine such as aldrin, chlordane, DDT, dieldrin, heptachlor.

Offsite: Outside the area or site where a pesticide is being used.

Omnivore: An organism that obtains its food energy from both plants and animals.

Omnivorous: Feeding on both animal and vegetable substances.

Ootheca: The egg case of certain insects that contains multiple eggs in a protective shell.

OP: Organophosphorus pesticide. Pesticides containing phosphorus.

Operculate: A body process that suggests a lid.

Opisthosoma: The end section of an arachnid's body also known as the abdomen.

Oral Toxicity: Ability of a pesticide to cause injury or death when taken by orally by mouth.

Organic matter: Materials and debris that originated as living plants or animals.

Organic Pesticides: Pesticides that come from natural sources.

Organism: A living thing that has the ability to act or function independently.

Organochlorine compounds: Organic compound containing at least one covalently bonded chlorine atom. There are numerous derivatives. They are controversial because of the effects of these compounds on the environment. Examples of organochlorine products include DDT, dicofol, heptachlor, endosulfan, chlordane, aldrin, dieldrin, endrin, mirex, and pentachlorophenol. Many of these agents have been banned.

Organophosphates: An insecticide that interferes with an insect's nervous system resulting in the disruption of vital nerve impulses this either kills the insect or interferes with the insects ability to function normally. Some nerve agents used in warfare have the same effect on humans examples being sarin, tabun, soman and VX. Multiple exposure to organophosphates inflates the toxicity.

Organophosphates: An insecticide that interferes with an insect's nervous system. Synthetic organic compounds that disrupt the nervous system.

Orifice: An opening or vent.

Ornithodoros: Part of the soft-bodied tick family.

Ostiole: A small bodily aperture or orifice.

Overhang: Outward projecting eave-soffit area of a roof.

Overhang: That portion of the roof structure that extends beyond the exterior walls of a building.

Ovicide: An acaricide or insecticide that kills the eggs of mites or insects.

Oviparous: Producing eggs which are hatched outside the body of the female.

Ovipositing: To lay eggs.

Oviposition: The laying or depositing of eggs.

Ovipositor: Egg-laying tubular structure at the end of the abdomen in many female insects.

Ovisac: A waxen sac into which eggs are laid. A sac containing an ovum.

P

Padding: A material installed under carpet for added comfort.

Palp: A segmented structure arising on the maxilla or labium which have a sensory function.

Palps: A pair of appendages ib arachnids located behind the fangs and used in sensing touch.

Papilla: A small bulge concerned with taste, touch, or smell.

Parabollically: With two or more non-parallel sides.

Parapet: A low wall along the edge of a roof or balcony, A wall placed at the edge of a roof to prevent people from falling off.

Parasite: An animal or plant that lives in or on a host.

Parthenogenesis: Egg development without fertilization. Process in which an unfertilized egg develops into a new individual; common among insects. Conception without fertilization by a male of a species.

Particle Board: Plywood substitute made of course sawdust that is mixed with resin and pressed into sheets.

Partition: A wall that subdivides spaces.

Pathogen: Any disease-producing agent resulting in the decline in health or even death of the host.

Patio: A paved backyard area.

Pellet: Hard form of faeces produced by some termites and insects.

Penetrant: Anything that helps a pesticide penetrate a surface or the shell or skin of an organism.

Penis: The male organ of copulation.

Pergola: An outdoor structure with climbing plants and an open roof.

Perimeter Spraying: The chemical treatment to the exterior perimeter of a structure.

Peristent Organic Pollutant: (POP's) Chemicals which do not break down easily in the environment.

Permeability: The ease with which water penetrates a material or other substance.

Permethrin: A synthetic chemical used as an insecticide, acaricide, and insect repellent which is highly toxic to cats and fish but generally has a low mammalian toxicity and is poorly absorbed by skin.

Personal Protective Equipment (PPE): Refers to protective equipment and clothing such as overalls, helmets and goggles, to protect the wearer's body from injury.

Pest animals: Wild animals that have a negative impact on the environment, agriculture or human activity

Pest Management: The manipulation and control of a pest or environment so as to reduce insect activity in a certain area.

Pest: Any unwanted, undesirable or destructive insect or animal.

Pesticide Resistance: Tolerance or immunity in a target pest to the pesticides being used to control it. Often this is the result of over exposure to sub lethal doses of the same control agent.

Pesticide: A chemical used to kill pests species.

Petiole: The slender stalk that attaches a wasp nest to a structure.

Petroleum-based: Made from petroleum products. Examples, refined oil and kerosene.

pH: The measure of acidity or alkalinity.

Phenotype: What an organism looks like as a consequence of the interaction of its genotype and the environment.

Pheromone: A chemical substance released by some insects to influence the behaviour of other insects of the same species.

Phragmotic: Any method an animal can use to defend itself in its nest or burrow by using its own body as a barrier. An example being the large, heavily sclerotised, head of the soldier caste of some termite species that can be used to block entrances.

Physical Pest Control: Barriers made from metal, mesh or granulated substances that do not allow termites to pass through.

Physogastric: Describes the distended abdomen of a mature termite queen.

Phytotoxicity: Harmful or lethal to plants. The ability of a product to damage plants.

pi: (π): 3.14159265+. The ratio of the circumference of a circle to its diameter.

Pier: A column of masonry used to support other structural members.

Pipe: When mature trees decay from the inside the hollow running up the length of the tree trunk is called a pipe.

Piscivore: An animal that feeds on fish.

Pitch: The incline slope of a roof.

Pith: The soft spongelike central cylinder core occurring in the centre of a tree trunk or branch.

Plant Activator: A substance that activates a plants defence mechanism against pests or diseases.

Polycalic: A distributed nest system which usually has a central nest and associated satellite nests.

Polyethism: The change of behaviour that occurs as a termite matures.

Polymorphic: A species having several forms independent of the variations of sex.

Polyphagous: Feeding on a variety of plants and or animals.

POPs: Persistent Organic Pollutants. Chemicals which do not break down easily in the environment.

Population: A group of organisms of the same species inhabiting a given area.

Porous Surfaces: Able to absorb fluids. Surfaces that have tiny openings or pores that allow a pesticide to be absorbed.

Potable: Drinkable, as in safe water.

Powder Post Borer: A wood borer that attacks some hardwoods. Also known as lyctid borer.

PPB: Parts per billion.

PPE: Personal Protective Equipment. Refers to protective equipment and clothing such as overalls, helmets and goggles, to protect the wearer's body from injury.

PPI: Pre Puchase Inspection. A Pre-Purchase Inspection checks a house or unit for defects, existing problems and potential problems, before you buy.

Pre Puchase Inspection: PPI. Pre Puchase Inspection. A Pre-Purchase Inspection checks a house or unit for defects, existing problems and potential problems, before you buy.

Pre-Baiting: Allowing a target species to become accustomed to traps or food before setting the traps or offering baited food. Process of getting rodents familiar with traps prior to setting the traps.

Predator: An organism, insect or animal that survives by preying on other animals that are either smaller or weaker than themselves.

Prehensile: Adapted for grasping.

Preservative: A chemical added to timber to protect against decay, insects or decomposition

Presoldier: The last larval stage before the moult to soldier.

Pressure Relief Valve (PRV): A device which is designed to release any high pressure in a tank to prevent explosions.

Pressure-treated Wood: Lumber that has been saturated with a preservative under pressure to enhance absorption of preservative.

Prevailing Visibility: The greatest visibility met or exceeded throughout at least half of the horizon.

Prevailing Wind: The wind that is most common in a specific area.

Primary Reproductive's: The main reproductive female in a colony. See Queen.

Prime: Fill with priming liquid. The creation of a partial vacuum inside the pump casing, which allows water to flow into the pump.

Proboscis: A long flexible snout in the form of a tube usually used for feeding or touch.

Proctodeal: Anus to mouth feeding. Proctodeal feeding allows worker termites to transfer food to other castes.

Prosoma: Called the *prosoma* in some groups. The first body section of arachnids. Also known as the cephalothorax.

Prothorax: The foremost of the three segments in the thorax of an insect, and bears the first pair of legs. The 1st or anterior thoracic segment.

Protozoa: One-celled animals, the smallest of all animals and which can only be seen under a microscope. They breathe, move, and reproduce like multi-celled animals. Termites live on cellulose, mostly from the dead wood they chew, but they depend on protozoa in their gut to provide the enzymes that can digest the wood.

Pupa: An insect in the inactive stage of development and usually a non-feeding and inactive stage. The third stage, between the larval and the adult stage.

Pupal: Post larval stage. It is during the time of pupation that the adult structures of the insect are formed.

Pupate: To develop into a pupa.

Purlin: Horizontal structural member in a roof usually at right angles to the pitch of the roof

Pyrethrins/Pyrethrum: Botanically derived organic compounds that disrupt the nervous system. Pyrethrins are much more toxic to pests than mammals. Mammals are able to breakdown pyrethrins into less toxic chemicals, which are then excreted

Pyrethroids: Compound usually used to control some common pests. A pyrethroid is a synthetic chemical compound that closely copies the natural chemical pyrethrins produced by the flowers of pyrethrums. Pyrethroids are less acutely toxic than carbamates and organophosphates.

Pyrethrum: Botanically derived organic compounds that disrupt the nervous system.

Q

Queen: The primary reproductive female in a colony.

R

Race: Genetically distinct populations of a geographic region.

Radial: Arranged or having parts arranged like rays.

Rafter: Roof structural members that slope downwards to the eaves.

Rain: Water falling in drops from vapour condensed in the atmosphere.

Ranch Home: Long and low home on one floor with basement and an attached garage

Raptorial: Adapted for seizing and grasping prey.

Ream: Used for measuring paper. A ream is sometimes 480 sheets but more often 500 sheets.

Recharge: To refill a reticulation system after the efficacy of the last refill wanes.

Recruitment: The action of termites to get others to join them in a certain activity.

Rectal: Relating to the rectum.

Rectum: The posterior expanded part of the hindgut.

Release: Intentional or accidental discharge of a pesticide into the environment.
Relief valve: A device designed to open if it detects excess pressure.

Render: Cement or plaster applied to brick or masonry walls.

Repellant: Any substance that repels insects or animals.

Replacement Reproductive: A reproductive which takes the place of a lost primary reproductive.

Reproductive: An insect whose only job is to reproduce, for example a termite queen.

Reproductive: Capable of reproduction.

Residual Action: The ability of a chemical to work effectively after the treatment or application has been completed. See residual Insecticide.

Residual Insecticide: Products that can kill or affect a target pest after application. External conditions such as wind, rain, and humidity may effect length of residual action.

Residual Insecticides/Pesticides: Insecticides that remain persistent, and can kill or affect a target pest over a long period of time. External conditions such as wind, rain, and humidity may effect length of residual action.

Residue: Matter that remains after something has been removed. Pesticide that remains active after treatment has been completed.

Resistance: The capacity of an insect or organism to resist the effects of a pesticide.

Restricted-Use Pesticides: Pesticides that only can be used or sold to accredited personnel.

Retaining wall: A structure that holds back a slope and prevents erosion. Typically a wall made of stone, concrete or wood with a higher level of soil on one side than the other.

Rinsate: Pesticide contaminated water that results from cleaning or rinsing out pesticide containers or equipment.

Rodenticide: A pesticide or any chemical used to kill rodents.

Roof Batten: Small timbers fixed to the top of rafters used to provide the fixing point for roofing sheet or roof tiles.

Roof Void Access: An opening that is placed in the dry-walled ceiling of a home providing access to the attic/roof void.

Roof Void Ventilators: Openings provided to ventilate or remove hot air from an attic/roof void space.
Roof Void: The area above the ceiling on the uppermost level of a building and below the roof itself. Also known as the attic.

Rostrum: Beak like projection of the anterior part of the head of certain insects. The needle like mouthpart of sucking insects such a bed bugs.

Runoff: Surplus liquid exceeding the limit or capacity. Run off as waste.

S

Saline: Containing salt.

Salinity: The concentration of salt in a solution.

Saliva: A fluid secreted by the salivary glands and mucous glands of the mouth which has digestive and other properties.

Salmonella: A pathogen that can be spread by roaches or rodents.

Sanitation: Part of an effective IPM plan. Sanitation involves routine maintenance of a site, regular scheduled cleaning, the removal of clutter and food sources, and the removal or destruction of existing or potential pest harbourages.

Sapwood: The outer layers of a tree which are still living and contain nutrients.

Sarking: A reflective foil laminate that is installed inside roofs. It has many benefits including weather proofing, insulation and reduction of dust and sound. Sarking will also prevent a build up of debris in a roof void that can enter through gaps in roof tiles.

Satellite Nest: In relation to termites, an additional nest which supplements the central nest.

SC: Suspension concentrates.

Scale: A thin flake of dead epidermis shed from the surface of the skin.

Scavenger: An animal that feeds on refuse and other decaying organic matter.

Scientific Name: Internationally recognised name given to organisms.

Sclerite: Hard plate or element of the exoskeleton of some arthropods.

Security Fence: A strong form of fencing which provides security from trespassers.

Semi-Attached: The end units on Row or Town homes.

Semi-Detached: Two homes built either side of a common centre wall.

Sensitive Areas: Sites that are particularly vulnerable to the effects of chemicals or pesticides. These could be schools, childcare centres, food preparation areas etc.

Sensitive: Particularly vulnerable, responsive to harm from pesticide exposure.

Serrate: Toothed like a saw.

Setae: A stiff hair or bristle. The sensitive hairs on the body of some insects.

Shelter Tube: A type of termite gallery that is built over, rather than through a substrate. Made with soil, faeces, saliva and carton.

Side Split: The longest aspect faces the road and the stairways are at the sides of the house.

Sill: The bottom member of a door or window frame.

Skillion: A sloping roof without a ridge or peak.

Skirting Boards: A trim board placed against the wall around the room next to the floor.

Skylight: Window located in a ceiling or roof.

Sleeper: One of the cross braces that support the rails on a railway track sometimes used as garden edging and attractive to termites.

Social Insects: Insects that live in organised groups.

Soffit: The area below the eaves and overhangs

Soil Sterilant: A substance preventing the growth of plants and/or animals in the soil.

Soldier: In termites this is the caste which has a sclerotised head and whose role is defensive.

Soluble: Able to be dissolved in another substance.

Solution: A mixture of two or more substances.

Solvent: A liquid substance capable of dissolving other substances.

Species: A classification of organisms into groups based on similarities of structure or origin. An organism that has the ability to interbreed and produce fertile offspring.

Specific Gravity: The ratio of the density of wood to the density of water at 4 C.

Specifications: A detailed description of criteria for a piece of work to be undertaken.

Spermatophore: A packet of sperm.

Spiniform: In the form of a spine.

Spinneret: Small tubular appendage on some spiders from which silk threads, webbing, are released.

Spinule: A minute spine.

Spiracle: A breathing orifice usually occurring on the third thoracic segment.

Split Level: A house built on two levels; a bi-level house.

Split Level: Homes incorporating different levels of floors linked by short stairways, can be two, three or four levels.

Splitting: Formation of a new colony by the splitting of an existing colony.

Spot Treatment: Application of a pesticide to a small area or to an exact area of activity.
Spp, Heterotermes ferox.

Stadium: The interval between insect moults, *stadia*.

Stage: A specific period in development.

Stagnant: Still, unmoving.

Steel Balustrade: A type of steel railing support often utilised around swimming pools or on balconies for safety.

Steel Fence Gates: A movable barrier, usually on hinges, closing an opening to a steel fence.

Steel Picket Fencing: A modern version of the traditional picket fencing, constructed for maximum strength and durability.

Steel Pool Fencing: Fencing for the perimeter of a swimming pool, constructed of galvanised steel.

Sterile: Unable to reproduce.

Stomach Poisons: Poisons that must be swallowed in order to kill the target species.

Stomodeal: Mouth to mouth feeding.

Strain: A genetically distinct group of individuals.

Strata title: The most common title associated with townhouses and home units. The subdivision of a property into lots and common property. A legal system by which individual units of real property, such as apartments, units or offices, may be owned separately.

Structural Pest Control: The control of pests in and around buildings, homes, offices, industrial sites and other building structures.

Structural Timber: Strong and dimensionally stable timber used for its strength.

Structural: Relates to building methods and design, how a building is constructed and able to remain standing.

Stucco: A plaster applied while soft to cover exterior walls or surfaces.

Stud: A vertical wood framing member.

Sub Floor: The area below a floor.

Subcontractor: Someone who enters into a subcontract with the primary contractor, a person, partnership or company who contracts with the client to carry out part of the works.
Subgenital: Below the genitalia.

Subspecies: A sub-group of a species usually separated by geographical isolation and differing size, colour etc.

Subterranean: Being or operating under the surface of the earth, as in termites.

Supplementary Reproductive's: A reproductive which takes the place of a lost primary reproductive.

Surface Water: Water on top of the earth's surface.

Surfactant: A chemical agent capable of reducing the surface tension of a liquid in which it is dissolved which allows the treatment to stick and spread more efficiently. Also referred to as wetting agents.

Surveyor: Someone who determines the boundaries and elevations of land or structures.

Suspended Ceiling: A ceiling system supported by hanging it from the overhead structural framing. The ceiling panels are built of lightweight acoustic material laid into the grid.

Suspension: A mixture in which fine particles are suspended in a fluid where they are supported by buoyancy.

Swarm: A nuptial flight of alates. How new colonies originate.

Symbiosis: A mutually helpful relationship. The relation between two different species of organisms that are interdependent; each gains benefit from the other.

Synergism: Is the combined effect of two or more active ingredients working together to produce an effect greater than the effect of each individual ingredient.

Synergist: A substance which improves the effectiveness of an active ingredient. An example being Pipernoyl butoxide which is used to increase the effectiveness of pyrethroid insecticides.

Synthetic Pesticide: A pesticide in which the active ingredient has been artificially manufactured. Man-made pesticides such as pyrethroids.

Synthetic Pyrethroids: A pyrethroid is a synthetic chemical compound similar to the natural chemical pyrethrins produced by the flowers of pyrethrums. Pyrethroids now constitute a major proportion of the synthetic insecticide market and are common in commercial products such as household insecticides.

Systematic Name: A name that fully defines a chemical compound and is derived using a set of rules.

Systemic Insecticide: Chemical absorbed into the system of a plant which renders its parts, the roots, stems and leaves poisonous to invading organisms.

The intention being that the plant or animal is toxic to pests that feed on it.

Systemic Pesticide: Chemical absorbed into the system of a plant which renders its parts, the roots, stems and leaves poisonous to invading organisms. The intention being that the plant or animal is toxic to pests that feed on it.

T

Target animals: Animals identified as subject to management under a wild animal management program.

Target Pest: The pest at which the control method is directed.

Tarsi: An insect's foot.

Taxonomy: The science of classification of animals and plants.

Technical Grade Material: The initial pure form of an active ingredient.

Temperature: A degree of hotness or coldness of a body or environment. Temperature can be measured in degrees on the Fahrenheit, Celsius, and Kelvin scales.

Tension: A state or condition of being stretched, strained or pulled by a force.

Tentorium: Endoskeleton of the head.

Terminal: Relating to an end or extremity.

Termite Shield: A metal sheet that is placed in the exterior walls of a house near ground level, usually under the sill, to prevent terminates from entering the house.
Termiticide: Any substance designed and used to kill termites.

Termitidae family: see Termites. See Common Pest Section.

Termitophile. Name given to any other animal or organism that lives with termites.

Terrace: An elevated outdoor area.

Terrestrial: Of or relating to or inhabiting the land as opposed to the sea or air.

Thermometer: A device used for measuring temperature.

Thorax: The middle region of the body of an arthropod between the head and the abdomen. This section contains muscles and bears the legs and wings.

Timber Pest Inspection: TPI. Should be carried out at least every year, but in some instances more frequently depending on a number of factors. TPI's should include all areas of a home incluing internal, sub floor, roof void, outbuildings, fences and gardens.

Timber: The wood of trees cut and prepared for use as building material.

Tolerance Level: The capacity to tolerate unfavourable environmental conditions.
Town Houses: Also known as Row houses. Two houses or more houses attached side by side sharing walls.

Toxicity: The capacity of a chemical to do harm to an organism. Measure of a pesticide's ability to cause acute, delayed or allergic effects.

TPI: Timber Pest Inspection. Should be carried out at least every year, but in some instances more frequently depending on a number of factors. TPI's should include all areas of a home incluing internal, sub floor, roof void, outbuildings, fences and gardens.

Trachea: Minute tubes which permeate the insect body and carry gases to and from the various organs.

Translocation: The movement of a substance within a system or organism.

Trapping: Method for physically capturing a pest species.

Traps: Devices that restrain pests.

Tray Ceiling: The ceiling line resembles an inverted tray.

Trim: The visible finishing work on the interior of a building.

Truncate: Ending abruptly.
Truss: A structural support, framework of beams, usually arranged in a triangular shape. Trusses are often used to support roofs and floors.

Tubular Steel Fencing: A steel barrier made from tube shaped bars.

Two Storey Home: Two complete floors.

Typhus: A highly contagious disease transmitted by body lice and characterized by skin rash and high fever. Spread by arthropod, lice, fleas, mites, bites.

U

Underfloor Crawl Space: The area under the ground floor flooring.

Unit: A self-contained dwelling unit in a multi-unit building.

Unsanitary: Dirty and contaminated.

Unseasoned Timber: Timber in which the average moisture content exceeds 25 %.

V

Vaulted Ceilings: These angle or arch up from the walls to the ceiling providing a high, spacious feeling.

Venation: The pattern of veins on the wings of termites.

Vendor: A person who offers a property for sale. Someone who promotes or exchanges goods or services for money.

Veneer: A thin layer of quality wood that is glued on top of other inferior woods for aesthetical purposes.

Veneer: A thin layer or sheet of wood.

Ventilation: The circulation and supply of fresh air to an area by natural or artificial means.

Vestibule: Entrance or reception room.

Vestigial: Poorly developed.

Virucide: A pesticide that is used to kill viruses in plants.

Viscosity: Resistance of a liquid to shear forces.
Void Injection: A treatment type that is used for hard to reach and inspect enclosed spaces or voids where insects may live, hide or travel. Flushing agents may reveal level of activity by making pest visible as they exit the void.

Volatile: Evaporating rapidly.

Volume: The amount of space occupied by a substance or object.

W

Wallboard: Composed of wood chips or shavings bonded together with resin and compressed into rigid sheets

Warm-blooded: Warm-blooded animals have constant body temperatures. Mammals and birds are warm-blooded animals.

Warranty: A statement that guarantees the material and workmanship of a product or service, usually limited by time.

Water-based Pesticides: Chemicals that use water as the only diluent.

Weatherboard: Boards that cover external surfaces to keep out rain and other environmental factors.

Weedkiller: See herbicide.

Weep Hole: Small opening at the bottom of a wall that allows moisture to escape from the void beneath a structure.

Wettable Powder: A Wettable Powder is an insecticide or pesticide comprised of the active ingredient in a finely ground state combined with wetting agents that are added to water to create an insecticide solution.

Wetting agent: see surfactant

White Ant: Outdated and incorrect term still used by some to describe termites.

Wild animals: Animals, both native and introduced, living and reproducing in the wild on land and in water

Wind Shear: Any sudden change in wind speed or direction.

Window Spider: see Black House Spider. See Common Pest Section.

Wing Buds: The pads located on an insect's thorax from which wings grow.

Wood Destructive Insects: Any insect or organism capable of causing damage to wooden members.

Wood Preservative: A pesticide and or fungicide that is used to treat wood to protect it from insects and fungi thus increasing its durability and resistance from being destroyed by insects or fungus.

Worker: Sterile member of a colony of social insects that forages for food and cares for the larvae as in termites. Workers are usually also responsible for maintenance of the nest.

Wrought Iron Gates: A barrier made from a form of iron, almost entirely free of carbon and having a decorative structure, which is readily forged and welded.

X

Xylem: The woody part of plants whose main function is to conduct water and dissolved mineral nutrients from the roots to other parts of the plant. Cellulose is needed as it is rigid and allows this pressure needed to transport the nutrients without causing the cells to loose their own structure. The primary food of termites.

Xylophagous: Feeding on woody plant tissues.

www.ingramcontent.com/pod-product-compliance
Lightning Source LLC
Chambersburg PA
CBHW071234170526
45165CB00003B/1090